José Epeko

Farming techniques

José Epeko

Farming techniques

ScienciaScripts

Cover image: www.ingimage.com

This book is a translation from the original published under ISBN 978-620-6-70977-0.

Publisher:
Sciencia Scripts
is a trademark of
Dodo Books Indian Ocean Ltd. and OmniScriptum S.R.L publishing group

120 High Road, East Finchley, London, N2 9ED, United Kingdom
Str. Armeneasca 28/1, office 1, Chisinau MD-2012, Republic of Moldova, Europe
Printed at: see last page
ISBN: 978-620-7-75341-3

FARMING TECHNIQUES

COURSE PLAN

O. INTRODUCTION

In the beginning, agriculture was just an art, summed up by gathering, hunting and collecting. When man became sedentary and the population increased, this forced the application of visa techniques, leading to a vertical and horizontal agricultural revolution.

This linear approach, often described as 'top-down', is based, on the one hand, on the (at least implicit) conviction that the experimental station is essentially the place where technical models for progress are developed, and, on the other hand, on a clear and strict division of roles between the various players: It was up to the researchers to design these new models, and it was up to the extensionists to find suitable ways of passing them on to the farmers, who were responsible for applying them.

The technical themes proposed can be sector-based or more or less global; they can range from a specific cropping operation to a complete cropping system (combining several plants, the order in which they are grown and the recommended technical itineraries for each). The aim is generally to provide a coherent set of themes. The term "technical package", which has now fallen into disuse, flourished during the years of the . The technical package is most often developed around the creation of a high-performance plant material to which are associated the cultivation techniques likely, under given soil and climatic conditions, to express its potential to the full, It is easy to see why the experimental station is the place where such models are developed, with performance assessed on the basis of a set of criteria in which yield is the main criterion. The most highly valued model is the intensive system, which provides the best possible approach to the productivity potential allowed by the plant material and the biophysical environment,

The farmer has to implement the recommended techniques, and to be perceived as an obstacle to the expression of the potential that the models of progress aim to bring out. The terms "vocation of the environment" and "human obstacles to development" are key expressions of a dated discourse, revealing a vision of agricultural development directed towards an overriding objective, that of "enhancing" the environment. Farmers are the agents, and it is considered that they can only be the beneficiaries.

The results that can be observed after the more or less satisfactory adoption of the technique by farmers. Agricultural research readily blames either the farmers themselves (considered to be resistant to progress, **incapable of mastering new techniques**, or prisoners **of social constraints**), or the poor extension methods adopted. For example, the simple technique of sowing food crops in rows was rejected long before by the farmers themselves.

Farming techniques must be applied according to the edaphic, climatic and economic conditions of each region; the opposite is possible if there is the capacity to modify this limiting factor. An understanding of ecology is essential.

Efficient techniques leading to good yields must obey the interaction of several factors. For example, a good choice of seedbed, sowing date, maintenance, product conservation, etc. must all work together.

O.1.Objectives

The aim of the agricultural technology component course is to provide L1 development students with a global vision of agriculture as a basic area of activity or the cornerstone of development.

Specifically, it will enable students to:

- Differentiate between traditional and modern farming techniques.
- Controlling the two environments in which the plant develops (soil and atmosphere).
- Understanding the components of agricultural yield.
- Know a few special techniques for propagating and producing plants.

0.2. Prerequisites

To facilitate understanding, students should have a basic knowledge of general agricultural concepts. This is why certain basic notions will be referred to in this course.

0.3. Relationship with other courses

This course is at the crossroads and is closely linked with many others, including General Agriculture, Plant Science, Soil Science, Climatology, Rural Sociology, Mathematics and Animal Science. Agricultural extension, ...

CHAPTER I: GENERAL INFORMATION

1.1 . The **origins of** agriculture

Agriculture in the broadest sense encompasses both plant and animal production. Agriculture involves the domestication and production of plants and animals useful to man. Domestication makes animal and plant species subservient to human objectives, including food and a source of income to meet socio-economic needs. Economic necessity (the inadequacy of hunting and gathering in the face of the demographic explosion) was probably the main factor in the birth of agriculture. Agriculture established a sedentary lifestyle. Above all, it changed socio-cultural conditions by developing a sedentary lifestyle to the detriment of the nomadic or transhuman systems of the time. Villages, towns, nations and empires were built on the basis of cultivated land. Feudalism, based on the possession of land by nobles and the servitude of peasants, is characteristic of the agricultural age. **It was through agriculture that man became sedentary, and sedentarism enabled man to develop.**

1.2 Defining some concepts.

1.2.1. Agriculture

There are several definitions of agriculture according to different authors and contexts. For details, please refer to the general agriculture course. Agriculture is defined as the art of producing living plant matter for human and animal consumption by tilling the soil (**Mazinga, 2019**).

In the broadest sense, agriculture can be defined as a set of techniques designed to develop and exploit plants (phytotechnics) and animals (zootechnics) rationally and economically.

In other words, agriculture consists of obtaining production (animal and/or plant) on a certain surface, within a given physical and socio-economic environment, by using production factors (**land, labour, capital goods,** etc.). In the context of this course, the term agriculture has a more restrictive meaning insofar as the focus will be on plants.

Originally, agriculture was more of an art than a technique. It relied more **on skill and know-how**. Its development and rationalisation were only possible when it was seen as a set of techniques based on fundamental knowledge developed in the physical, chemical, biological and human sciences. Farming is often presented as one of the finest professions in human endeavour (Prévost, 1999). In addition to the nurturing role it has always played in society, it is difficult to remain indifferent to this profession, which demands all the intellectual, moral and physical qualities of the human being, which allows us to live in nature, and which leaves a lot of room for individual initiative. Faced with developments in various fields, the farmer must be at once: an art, a technique and a science - in this case, agronomy.

1.2.2. An agricultural technician

He is someone who has mastered the techniques of agricultural production;

1.2.3. An agronomist: i.e. someone with the necessary scientific knowledge (agriculture is not a recipe-based profession that you just have to follow, because the working tool is the living environment and nature is too capricious to be stable):

- **A good manager, all the more so** as agriculture requires a lot of inputs (farming equipment, fertilisers, pesticides, etc.) and profitability margins are often narrow for many crops;
- **A good business manager** who knows how to manage his workforce (workers) and farm equipment;
- **A public relations man** who knows how to maintain good relations with the entire professional community (extension services, cooperatives, transporters, input suppliers, etc.);
- **A trader** who knows how to sell his products and earn a better income from them (agricultural products are subject to the law of the market, i.e. supply and demand);
- **A man of progress** who knows how to adapt to new economic, social and cultural contexts, and who is also capable of innovation. We can see in him a socio-economic evolution.

In the light of the above, we can see that farming is a complex and difficult profession. Sometimes the satisfactions it brings are not equal to the problems and worries the farmer encounters. It is therefore advisable to start production on a small scale and expand as production is mastered.

1.2.4 Agronomy is the study of the physical, chemical, biological and economic problems posed by the practice of agriculture. The agronomist is the person who devotes himself to the agronomic sciences. They seek to discover the laws of agricultural production and establish agricultural theories. Agronomists need to know the mechanisms of agricultural production in detail. Very often, developing solutions to agricultural problems requires a mind capable of integrating numerous components, and most of the time requires a multi-disciplinary approach permeated by the different agronomic sciences. **The role of the agronomist is to help optimise agricultural production with a view to promoting food security and economic development.**

Among other things, the agronomist must :

- Orient its activities towards professionalism, in particular by developing a spirit of initiative and competition
- Participate in the identification of problems in the agricultural sector and- Contribute to the gradual accumulation of the technical references needed to resolve the problems that arise.

1.2.5 Agronomy and agriculture: Agronomy and agriculture are two different fields. Agronomy brings together a range of scientific disciplines focused on the acquisition of knowledge. Agriculture, on the other hand, is a finalised activity involving human action on the environment. On another level, agriculture cannot be considered as a simple application of agronomy, given the many other aspects that govern its operation. Although agronomic research is concerned with real-life agricultural situations, the advances in knowledge acquired in agronomy through experimental approaches do not always translate into similar advances in agricultural practices. In particular, agronomy makes it possible to : - develop diagnostic approaches in production environments; - develop and transfer new techniques.

Agronomy is a science that brings together several disciplines. - Plant production- Animal production- Plant fertilisation, animal nutrition and human nutrition- Crop protection and defence- Soil science- Water management- Biotechnology- Agricultural machinery- Agricultural economics- Food industry- Environmental science and engineering- Waste recovery and recycling- Ecology and the environment, etc. Many fundamental sciences contribute to the development of agriculture, including:- mathematics- statistics- physiology- biochemistry (science dealing with the problems of life, the constitution of living matter and the chemical reactions that take place in living organisms);- microbiology- biology- botany (the science of plants);- genetics- pedology (the study of existing soils, taking into account their evolution in the past and their future development trends).

1.2.6 Food safety

The concept of food security refers to the availability of and access to food in sufficient quantity and quality. It helps to combat undernourishment and malnutrition and has four dimensions:

1. Availability (domestic production, import capacity, storage capacity and food aid).

2. Access (depends on purchasing power and available infrastructure)

3. Stability (in terms of infrastructure as well as climatic and political stability)

4. Health, quality (hygiene, mainly access to water).

Food security goes beyond the notion of food self-sufficiency. Food security is not necessarily achieved when the food supply is sufficient. We need to be able to answer questions such as "who produces the food", "who has access to the information needed for agricultural production", "who has sufficient purchasing power to acquire the food" and finally, "who has sufficient purchasing power to acquire the information needed for good production". Different options are possible for increasing agricultural production, through the adoption of specific agricultural production systems: vertical and horizontal revolution, in deficit countries, which can become self-sufficient.

Food self-sufficiency The ability of a country to meet the food needs of its people through production alone.

1.2 Some branches of agriculture

- **Arboriculture**: refers to the activity of growing trees, usually fruit trees, to produce fruit. (Mango, citrus, rambutan, etc.).
- **Sylviculture** (from the Latin silvæ, woods and forests, and Cultura, culture) is the rational exploitation of forest trees. It consists of the conservation, maintenance and natural or artificial regeneration (reforestation) of forest areas. Forests maintained in this way are generally used for commercial purposes (furniture, paper pulp). However, forestry is also involved in environmental protection (sustainable development) and landscaping.
- Market gardening: Market gardening is a branch of agricultural production that includes the technical, economic and food-related activities involved in obtaining vegetables for food and condiments. It involves growing vegetables and certain fruits, or even flowers, for food use intensively and professionally, with a view to marketing them. Market gardening is a type of intensive agriculture, which aims to maximise soil use and produce in very short time cycles. On the other hand, it requires considerable resources (irrigation network, greenhouses, fertile soil, etc.) and a large workforce, as mechanisation is difficult to implement for this type of crop.
- **Horticulture** refers to the branch of agriculture devoted to the production of garden plants. The term is derived from the Latin *hortus*, garden. Horticulture is opposed to agriculture in the sense of open field cultivation (*ager*, the field). Its field of activity includes food crops: vegetables (market

gardening), fruit trees, ornamental plants: cut flowers, perennials, ornamental shrubs and trees, and industrial crops: aromatic, spice or perfume plants (vanilla, cinnamon, etc.).

- **Agrology** (the study of soil properties in relation to crop production).

CHAPTER II: TRADITIONAL AND MODERN AGRICULTURE

The remarkable difference between these two concepts is that, starting with traditional agriculture, certain ancestral techniques have been improved to achieve modern agriculture. Certain methods are still effective between the two terms (points in common), such as crop association, the difference between which lies solely in feasibility (choice of crops, taking into account incompatibilities and compatibility, the right time to do it, the spacing adopted, etc.).

2.1 TRADITIONAL AGRICULTURE

Traditional agriculture is a sub-branch of agriculture characterised by low use of technology, based on mixed farming, with production primarily for the farmer's own consumption

Here are some of the characteristics of this type of agriculture:

- Use of unimproved seed, often attacked, often obtained from the previous harvest, sometimes degenerated seed.
- Incineration is the only technique used to open up the land, and is often referred to as shifting agriculture (agriculture itinérante sur brulis), or nomadic cultivation in search of fertility.
- Low density: as a result of haphazard sowing, failure to respect plant spacing and the presence of tree stumps and trunks, which reduce the area available for cultivation.
- This is extensive farming with a low yield per hectare.
- Use of farming tools (machete, spade, hoe, etc.).
- Fallow land is the only effective cultivation technique for restoring fertility, but organic fertilisers (manure) and other fertilisers such as ash and guanos are also used.
- Rain is the main source of water (rain-fed agriculture).
- A polyculture with the most preferred species installed in a disorderly fashion.
- The non-mechanisation of agriculture and the use of synthetic pesticides to protect crops.

In underdeveloped countries, it is thanks to this agriculture that the population is spared undernourishment. It is this agriculture that is the cornerstone for scientific research in agronomy.

2.1.1 ITINERANT BUSH FARMING

Slash-and-burn agriculture is any agrarian system in which the land is opened up by burning, and the land is cultivated discontinuously, with a fallow period longer than the duration of cultivation (Conklin, 1957). Here we see cultural nomadism in search of fertility.

Slash-and-burn agriculture comprises several phases:

(I)Clearing of a portion of forest (scrub clearance and/or felling) ;

(ii) burning of plant debris ;

(iii) cultivation of the land for a generally short period ;

(iv) set aside for a generally long period.

Slash-and-burn agriculture alternates between periods of cultivation and forest regeneration. It is therefore not a permanent form of agriculture.

Anglo-Saxons use several synonyms. Among the most common, **slash-and-burn agriculture** refers to the removal of forest cover and fertilisation by fire, while **shifting cultivation**, **swidden agriculture**, **shifting agriculture** and **itinerant agriculture** refer to the mobility - in space and time - of the cultivated portion of land.

Figure 1: Opening up the land by burning it.

Figure 2: Slash-and-burn agriculture.

In the intertropical zone, it is a good adaptation to the climatic and soil conditions of the rainforest, enabling the indigenous people to benefit from the fertility of the forest soils and the completely decomposed humus, opening up the land less expensively,

This method has many drawbacks, including deforestation, a rapid decline in soil fertility, soil sensitivity to erosion and global warming. Nowadays, with demographic pressure, this technique of shifting cultivation over burnt land is to be avoided because of its damaging effects. Farmers are being called upon to make plant production more sedentary, since arable land is shrinking steadily while the population is increasing exponentially.

2.1.2. EFFECTIVE TRADITIONAL TECHNIQUES.

Land development is characterised by the association and/or juxtaposition of agriculture and livestock farming. Farming systems are relatively complex, and the development that accompanies them varies according to the preponderance of agriculture over livestock (or vice versa) and the permanence of human settlements. A distinction is made between intensive and extensive systems.

2.1.2.1 TRADITIONAL INTENSIVE SYSTEMS

They are practised on the inhabited areas of the plateau, where family farms have access to the three types of land that are essential to the autonomy of each family (lowland, hillside and hilltop). Intensive mixed farming is combined with small-scale livestock rearing (sheep, goats, pigs, poultry) in a hedged landscape. The family farms (generally less than 3 ha) stretch along the slopes and combine a varied range of plants, from the summit to the valley.

The upper slopes are either pastureland for sheep and goats (*Pennisetum purpureum, Panicum maximum*) or food fields (temporary or permanent) where groundnuts are grown alongside maize, sweet potatoes, beans, yams and potato peas. Where densities are low, fire is used to clear land and renew fodder.

2.1.2.2 The Techniques For Maintaining Soil Fertility Are Versatile

Fallowing allows the natural replenishment of soil nutrients. On intensive food crop fields, it is a short fallow between cropping seasons, whereas on hilltop groundnut fields, the fallow is annual or multi-annual.

- **The integration of livestock farming with cultivation**: plots of land left fallow are grazed by sheep and goats, who make use of the crop residues and deposit their manure. The pig pens are regularly moved around the dwellings and the freed-up land is cultivated.

Figure 3: Integration of livestock farming with agriculture in the strict sense.

Burying organic matter under the ridges: crop residues, livestock manure, domestic waste and ash and any organic matter that could enrich the soil are stored in the furrows and covered with soil when the fields are prepared. However, as not all the organic manure is transformed during the cropping season, ploughing re-exposes un-humified waste on the surface, which partially protects the soil.

Biomass recycling: this is particularly effective given the regular alternation between ridges and furrows. While the ridges support the crops, the furrows receive the domestic and weeding waste that will fertilise the future ridge. In this way, each growing season, one part of the soil is less stressed than the other and is replenished to support crops the following season.

➢ **The practice of "ecobuage":** this consists of piling up the grasses pulled from the plots, covering them with earth and then setting fire to them from a hole in the side. The slow combustion preserves all the ashes from the burning, protects them from rainwater and helps to fertilise the soil.

Figure :4 Wood burning

➢ **There are various methods of erosion control:**

Combining several crops on the same ridge: this ensures ridge stability, good soil cover and reduces erosion. So it's easy to see why large ridges laid perpendicular to the slope are so effective at resisting run-off.

The practice of two cropping seasons: limited to food plots, this ensures permanent soil cover, especially when the first season's crops are partly present on the fields.

Combining trees with crops: these fruit or forest trees provide the necessary shade for certain crops, slowing wind speeds and preserving soil moisture. Their litter protects the soil from the impact of raindrops and slows erosion.

Keeping crop residues on the fields: these include maize stalks left standing, groundnut tops left in the furrows, etc., which mulch the soil and protect it from the high levels of sunlight in the dry season and the harmful effects of run-off. Some stalks are used to support yams planted later.

The size and layout of the ridges on the cultivated plots: these vary according to topographical position, crop type and soil thickness (only the length of the ridge is sometimes dictated by the size of the cultivated plot).

2.1.3 OBTAINING PROPAGATION MATERIALS

Propagation material is often harvested from the previous crop, and farmers refer to their seed selection criteria. Seeds are either stored in granaries or sown directly.

Choice of land: farmers refer to certain fertility indicator plants, or the length of the fallow period is used.

2.2 MODERN AGRICULTURE

Modern agriculture (conventional, which has become productivist and is also known as agribusiness) can be intensive or extensive: it uses machinery, improved seeds, provides high production, reduces the cost of labour or working time.

The characteristics of modern agriculture :

Sophisticated machinery is used at every stage of the growing cycle.

- Chemicals to boost production;
- Modern techniques, sowing in rows, improved seeds;
- Fertilisation: Fertiliser and soil improver spreaders provide the soil with the mineral or organic elements needed for plant growth.

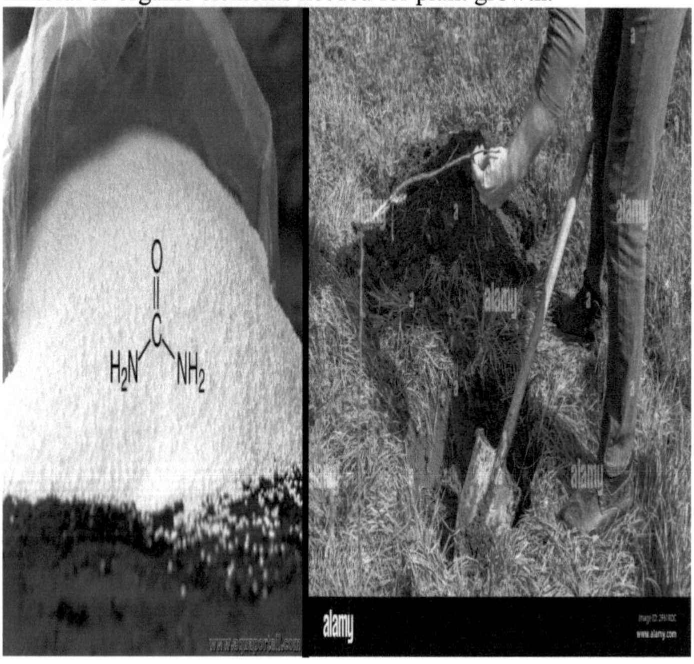

Figure :5 Fertilisers (mineral and organic)

- **Phytopharmaceutical protection**: to combat crop pests (diseases, weeds, insects, fungi), the AOS uses pesticides applied with a sprayer or deposited directly on the seed, or mechanical control.
- **Soil preparation:** various types of equipment are used to work the soil: ploughs, stubble cultivators, harrows, ridging machines, etc.

Figure 6: The ridge-forming machine.

- **Seeding or planting**: seeders and planters

Figure 7: The single-seed drill and planters

Perennial crops are often first sown in nurseries in polybags,

Figure 8: Polybags

- **Harvesting:** each crop has its own harvesting machine.

Figure 9: The rice harvester.

- **The use of special vegetative propagation techniques:** grafting, layering, PIF method, micro propagation, etc.

Figure 10: Grafting (rubber tree cultivation).

- **The disadvantages of modern agriculture :**

Pollution, by dumping large quantities of manure, chemicals and antibiotics into water sources. This puts both ecosystems and human health at risk.

The high cost of production due to the purchase of many inputs and specialised techniques.

To overcome the disadvantages of modern agriculture, we need to use methods that increase yields while conserving the environment: agroecology, organic farming, sustainable agriculture, integrated farming, etc.

a) Agroecology uses ecological theory to study and manage agricultural systems in order to improve their productivity and efficiency in conserving natural resources. This holistic approach to the development of agricultural and food systems draws on a wide range of technologies, practices and innovations, including local and traditional knowledge and modern science. By understanding and working on the interactions between plants, animals, humans and the environment within agricultural systems, agroecology encompasses multiple dimensions of the food system, including ecological, economic and social dimensions.

b) Organic farming is a production method free from synthetic chemicals, genetically modified organisms, growth regulators and feed additives. It also emphasises a holistic approach to farm management in which rotation cycles and animals are an integral part of the system. Soil fertility is the cornerstone of organic management. As organic farmers do not use synthetic nutrients to restore degraded soils, they must focus on building and maintaining soil fertility primarily through their basic farming practices.

c) Conservation agriculture: conservation agriculture practices have significantly improved soil conditions, reduced land degradation and boosted yields in many parts of the world through the implementation of three principles:

- Minimal soil disturbance ;
- Permanent ground cover ;
- Crop rotation.

For a farming system to be truly sustainable, it is essential that the loss of organic matter neverexceeds the soil formation rate (the mineralisation coefficient must be less than or equal to the humification coefficient). In most agricultural ecosystems, it is impossible to respect this parameter when the soil is subjected to mechanical disturbance. This is a key element of Conservation Agriculture, limiting the use of mechanical tillage (or soil preparation) in the farming process.

Scarification and localised ploughing can also be used (just where the crop is to be planted).

Zero tillage is one of the techniques used in conservation agriculture. The main aim of this technique is to maintain a permanent or semi-permanent biological soil cover (e.g. a standing crop or mulch) that protects the soil from the sun's rays, rain and wind and allows soil micro-organisms and fauna to work the soil and maintain the balance of nutrients.

d)Agroforestry systems include both traditional and modern land use systems where trees are managed alongside crops and livestock in agricultural structures. Combining the management of trees, crops and livestock mitigates environmental risk, creates permanent soil cover to combat erosion, minimises flood damage and contributes to water storage, benefiting crops and pasture. We can distinguish between the Agro-pastoral system, the Sylvoagropastoral system, etc.

CHAPTER III: ENVIRONMENTS FOR PLANT DEVELOPMENT

Plants live in two environments, the soil and the atmosphere. The soil provides the plant with nutrients, in this case water and mineral salts (macroelements and microelements in ionic form) absorbed by the absorbing hairs. The atmosphere supplies carbon dioxide (CO2) and water, which is essential for photosynthesis using light energy. It is also the source of nitrogen and many other gases and elements essential to the plant's metabolism, including climatic factors, oxygen, etc.

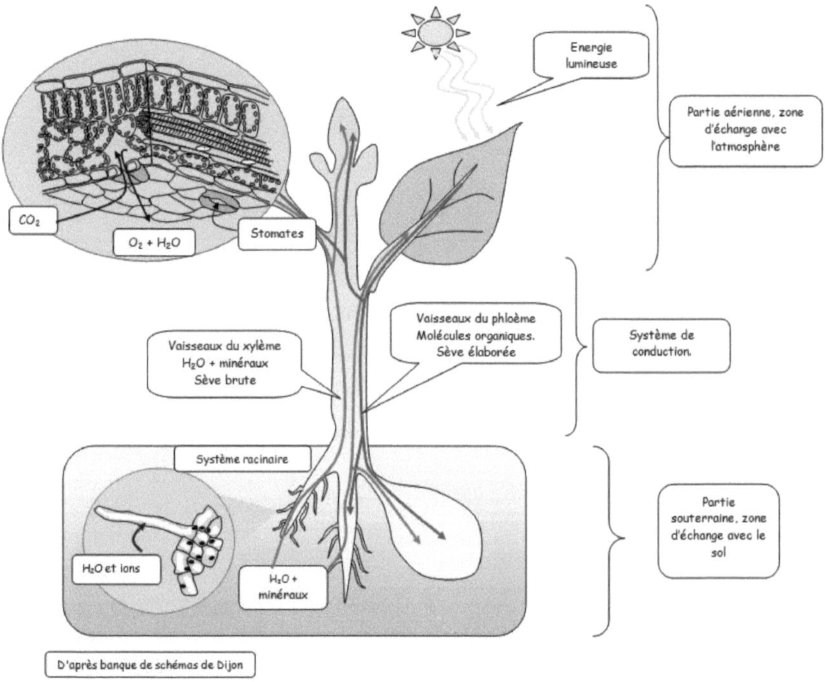

Figure 11: Living environments for plants (soil and atmosphere).

3.1. THE SOIL

The quality of the substrate in which cultivated plants grow has a considerable influence on their productivity potential. Soil properties can vary greatly depending on climate, position in the landscape, proximity to watercourses, the nature of the parent rock and human intervention. In general, the chemical fertility of tropical soils is much lower than that of soils in temperate regions, as they have undergone much more intense weathering processes associated with high temperatures and heavy rainfall. As a result, they are generally much poorer in nutrients, and consequently more acidic and more acidifiable. They contain virtually only kaolinite as a type of clay, and the humus content plays a much greater role in their overall fertility level than in temperate regions.

Micronutrient deficiencies in the soil cause malnutrition and micronutrient deficiencies in people, as the agricultural produce grown on these soils tends to lack the nutrients needed to combat invisible nutritional deficiencies." **M.....**

The soil is the plant's source of nourishment, and a well-balanced soil will make the plant stronger and more beautiful. It will be better able to resist climatic and parasitic aggressions.

Soil functions (soil cover) perform four groups of essential functions in relation to human needs and health:

- Biological functions The soil is partially or completely home to many animal and plant species, and many biological cycles include the soil. Moreover, the soil's biological activity is essential to its construction, functioning and fertility: aggregation, porosity, availability of nutrients, etc. Soil does not exist without abundant and diverse biological activity.
- Food functions The soil produces and contains all the elements necessary for life; it accumulates most of these elements and makes them available to plants and animals. It acts like a larder, more or less large and more or less full. Much of what plants absorb and breathe comes from the soil, and plants use, directly or indirectly, the entire depth of the soil, up to several metres below the surface.
- Exchange and filter functions Soil is a porous medium through which water and gas flows constantly. Water from wells and springs has already passed through the soil; the porosity of the soil influences its supply. Soil also acts as a filter, transforming water as it passes through it. The chemical and biological quality of water depends on soil properties.

Soil health has been defined as the soil's ability to function as a living system. Healthy soils maintain a multitude of soil organisms that help control plant diseases, weeds and insect pests, form beneficial symbiotic associations with plant roots, recycle essential plant nutrients, improve soil structure with positive effects on the water and nutrient holding capacity of the soil, and ultimately improve agricultural production. A healthy soil also helps to mitigate climate change by maintaining or increasing its carbon content.

Physical properties (porosity, structure, texture, hydromorphy), chemical properties (pH, aluminium toxicity, chemical element content, etc.) and biological properties must all be taken into account in relation to crop requirements.) Within this framework, the soil can be selected according to the requirements of each crop, or the soil properties can be improved according to the requirements of each crop.

For example, leafy vegetables and certain cereals require nitrogen-rich soil with a high organic matter content. These plants are generally grown at the head of the rotation.

Figure 12: Soil enriched by green manures for leafy vegetables.

Geocarpic plants require light soil to facilitate the development of their storage organs. This group includes tuber-rooted plants, tuberous plants (yams, potatoes, sweet potatoes, taro), groundnuts, etc. If the soil is heavy, it needs to be loosened by tilling. This is why sweet potatoes and manioc are planted on mounds.

Root and tuber crops require potassium-rich soil, while legumes require calcium-rich soil.

In nitrogen-poor soil, legumes are able to fix atmospheric nitrogen in symbiotic association with bacteria. The plant supplies the bacteria with photosynthate and the bacteria in turn supply atmospheric nitrogen for their metabolism.

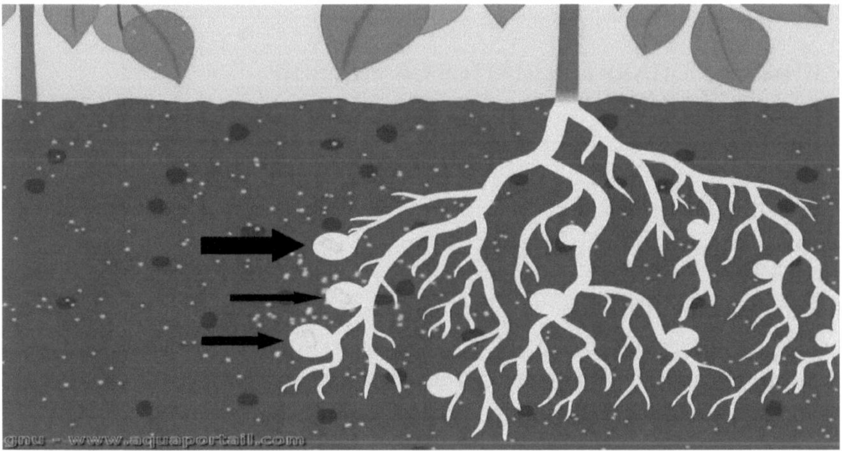

Figure 12: Symbiotic association.

In hydromorphic soil, plants that can withstand the presence of water should be planted, particularly leafy vegetables. Waterlogged soils are also well suited to species such as banana, aquatic species adapting to this phenomenon is explained by the adaptation of the root system, which is transverse.

Figure 13: Hydromorphic soil

For crops that cannot tolerate waterlogged soils, the soil must necessarily be shaped to increase its depth or thickness, which can be exploited by the roots. These include

- The mound ;
- Ridges ;

- Drainage ;

CHEMICAL CHARACTERISTICS OF THE SOIL

Mineral reserves

Plants draw many mineral substances essential to their metabolism from the soil. These substances are classified as :

Macroelements: These are the important elements that the plant needs in large quantities, for example: N, P, K, Ca, Mg, S, while the microelements are the elements that are just as important for the plant, but the plant needs them in small quantities. Fe, Zn, Co, Mn, B, Mo, Cu.

The soil content of each of these elements, their proportions and their availability to the plant are factors that influence plant growth. Deficiency or excess of any one of these elements manifests itself in a variety of specific symptoms, the effect of which is to reduce yields.

By exporting these elements, the plant impoverishes the soil, making it less and less able to support new crops. Maintaining high soil productivity therefore means restoring the elements exported by crops. Soil regeneration can be achieved either by leaving it fallow or by fertilising it.

Fallow land is the state of a parcel of land between the harvesting of one crop and the planting of the next. It is characterised, among other things, by its duration, the cultivation techniques applied to the land and the roles it fulfils. Fallow land is part of the succession of crops over time and therefore part of the cropping system. It can last from a few months to several years.

Previous effect (previous crop)

The previous effect of a crop is defined as the variation in the state of the environment between the beginning and the end of a crop or a fallow period, under the combined influence of the plant and the cultivation techniques, the whole being subject to the action of the climate. Understanding a previous effect does not mean establishing a statistical relationship between the yield of one crop and the yield of the crop that preceded it. It means analysing the changes in the physical, chemical and biological parameters of the plot under the effect of a crop.

As a result, we cannot define the preceding effect of a plant species: the changes in the environment induced by a cassava crop have as much, if not more, to do with the way in which the cassava was grown than with the fact that it is a *cassava* species. By definition, the previous effect is independent of the crop that follows. This does not mean that all crops will react in the same way to the environmental conditions linked to the previous crop, but that in order to understand the previous/following relationships we need to introduce a second concept, that of the sensitivity of the following crop.

Manuring

We distinguish between organic fertilisation, mineral fertilisation and the inclusion of enriching plants in the crop rotation.

Organic manure

It comes from the decomposition of plant and animal residues. It not only provides nutrients, but also organic matter, and plays an important role in the physical and biological properties of the soil. Losses are linked to its mineralisation, which is very rapid at high temperatures. This explains why tropical soils are generally poor in humus.

Organic matter is of considerable importance for both soil and plants, in that it :

- Improves rainwater infiltration;
- Increases water retention capacity and CEC; - Improves soil structure (cementing role).
- Reduces erosion ;
- Retards the leaching of nutrients, which it retains energetically;
- Increases biological activity in the soil, serving as a source of matter and energy for soil micro-organisms;
- Gradually releases fertilising elements, particularly NPK.

Ways of maintaining or even increasing the soil's organic matter content include :

- Organic fertiliser: one tonne of straw provides ¼ of its weight in humus ;
- Fertilisation: as a result of more vigorous plant growth ;
- irrigation, for the same reason;

- Incorporating a cereal into the rotation ;
- Grass fallow ;
- - non-incineration.

There are several types of organic fertiliser:

- **Manure: this is** a mixture of animal waste and **straw**,
- **Compost**: this is a mixture of decomposed plant waste,
- **Green manures**: these are plants that are sown so that they can be buried in the soil to decompose,
- **Liquid manure**: organic fertiliser made from liquid animal excrement.

The pH

Soil pH influences plant growth, in that it affects, among other things, the solubility, and therefore the availability to plants, of mineral elements. The optimum pH for most crops is around neutral, between 6 and 7.

In the humid tropics, the intense leaching of exchangeable bases (Ca^{2+}, Mg^{2+}, K^{+}), soils are generally acidic, with pH levels often below 5.5. Under these conditions, there are often deficiencies in Mo, Mg^{2+}, Ca^{2+} and P. In addition, acidity is accompanied by toxic doses ofAl^{3+}which prevent normal root growth in many plants, such as sugar cane.

In these soils, it is essential to raise the pH. The classic method for this is **liming**. This technique was already known 200 years B.C. It is carried out by applying lime ($Ca(OH)_2$), limestone ($CaCO_3$) or dolomite ($CaMg(CO_3)_2$. These applications not only correct the acidity, but also add Ca^{2+} (and Mg^{2+}). The amount to be applied varies according to the nature of the soil and the type of crop.

However, the pH can be reduced by using the sulphate form, for example ammonium sulphate. This operation is called **plastering**.

pH also affects microbial activity in the soil. In acid soil, most of the bacteria in the nitrogen cycle are not very active. This is true of nitrogen-fixing bacteria. This partly explains why less N is fixed in the tropics than in temperate regions; the effect of pH on microbial life also explains why certain fungal diseases only develop in acid soils, while others only proliferate in alkaline soils.

Biological characteristics of the soil: the soil is a living environment in which the micro- and macro-organisms that make it up play a part in its formation. For example, earthworms are considered to be the soil's engineers.

Figure 14: Biological characteristics of the soil

3.1 CLIMATE

Before looking at the climate, let's start by mentioning the climatic factors: precipitation, temperature, light and wind. Only temperature is a factor that cannot be easily manipulated, whereas the others are easy to handle. For example, in the event of water stress, irrigation is used.

In the tropics, it is generally not the temperature of the environment that is the most limiting factor for crop production, but the availability of water throughout the year. The year is divided into dry and rainy seasons. Generally speaking, near the equator, the climate is wet for most of the year. As you move away from the equator towards the tropics, there are increasingly long dry periods. In the absence of interference from large mountain ranges, there is a gradual shift from bimodal rainfall patterns (with two rainy seasons and two dry seasons) to unimodal patterns, with a single major rainy season and a single major dry season. The total amount of annual rainfall also tends to decrease away from the equator, and the distribution of rainfall tends to become less regular. Near the tropics, in the driest but sunniest parts of the tropical world, the availability of irrigation means that very high levels of production can be achieved. The ability to make up for a lack of rainfall (in terms of total quantity and distribution over the year) through irrigation is a factor that greatly limits production risks and maximises the efficiency of the production factors used. In the tropics, temperature is a limiting factor for crop production only in high-altitude areas (generally only during dry periods, when the climate is also the coolest).

The rule of thumb is to grow each crop in the right temperature environment.

Other climatic factors cfr the general agriculture course.

CHAPTER IV: PERFORMANCE COMPONENTS

4.1 Cultivation techniques for good yields

As well as the factors involved in agricultural production (land, labour and capital), we will be talking here about the technical itinerary to be followed to obtain a good yield.

Planning: farming begins on paper, where you have to plan the area, the workforce, the crop, the risks, the quantity of seed to be used, the feasibility and the timetable for each operation.

- **The choice of seed:** The choice of seed determines the success of a given crop, especially cross-pollinated plants. Research centres should be used every time to obtain good quality seed because hybrid vigour decreases with each generation. For self-pollinating plants, it is also preferable to turn to research centres to obtain improved seeds following the phenomenon of self-pollination and cleistogamy, even though there is always a rate of cross-pollination. In this case, the seeds must be harvested in the field for rigorous mass selection, as the loss of varietal characteristics is low compared with cross-pollinating species. The parameters to be checked are purity, germination capacity, cultivation value, etc.
- **Site preparation:** Site preparation consists of opening up the land (non-incineration, incineration, or an intermediate method) and working the soil. Tillage depends on the type of crop, the physical properties of the soil and the topography of the land.

Establishment: this stage includes direct sowing, planting and replanting. In the case of rain-fed agriculture, the main rule is to respect the agricultural calendar so that sowing and planting coincide with the rainy period.

Caring for the crop: Caring for the crop involves controlling its enemies and maintaining it in good conditions for production. When it comes to weed control, timing is everything, so it's best to intervene before the vulnerable stage; the stage at which weeds start to cause damage to the crop, so that direct control is possible. This reduces the frequency of maintenance and facilitates the normal growth of the plant, as intervening too late is a waste.

For the pest, we need to intervene at the economic threshold, i.e. the attack rate will reduce production. Preventive methods are always advisable, as they offer many advantages.

Depending on the different weeding methods used, we can mention : Ring weeding, **strip weeding**, **select weeding** and **general** or **clean weeding**.

Other maintenance operations include earthing-up, repotting, staking, trellising, pruning, etc. All these operations will be developed during practical work in the field.

Harvesting: When harvesting, the first aspect to take into account is the maturity of each plant. For plants with sequential flowering or in order to economise on crops whose harvesting time can be postponed, **harvesting is staggered** (as in the case of cassava, yam and beans).

Figure 15: Oil palm and bean harvest

Post-harvest operations: transport, storage, **drying....**

4.2 Performance components :

These parameters may or may not have a direct influence on crop yield, and they allow us to make correlations with yield, although they also depend on the nature of the useful product targeted by each crop.

a) Vegetative parameters :

We have those that are quantitative and qualitative.

- Germination rate ;
- Lift rate ;
- Diameter at the collar ;
- Final plant height ;
- The number of branches ;
- Number, colour and shape of leaves ;
- The number of nodes and the length of the internodes ;

- Colour and average length of petioles ;
- Leaf area

b) Generative parameters.

- The average number of flowers ;
- Flower abortion rate ;
- Average number of fruits ;
- Average fruit weight ;
- The number of tubers and tuberised roots ;
- The shape of tuberised T&R ;

c) Phytosanitary parameters.

- The impact ;

$$\textbf{Incidence}=\frac{number\ of\ diseased\ plants}{total\ number\ of\ plants}x100$$

- Severity: Level of severity expressed in relation to the score. It can be determined on the basis of symptoms or calculated.

$$\textbf{Severity}=\frac{quotation\ sum}{total\ number\ of\ plants}$$

Example of some rating scales :

- The case of MAM (African Cassava Mosaic).

1. No visible symptoms (apparent resistance)
2. Beginning of symptoms (chlorotic patches)
3. Mosaic spread over the entire leaf surface
4. 2/3 leaf distortion and general leaf size reduction
5. Severe mosaic, severe distortion of 4/5 of the lanceolate leaf.

- Baselle rust

1: No attack, 0% of the leaf surface ;

3: Weak attack, 1 to 25% of the leaf surface, i.e. ¼ ;

5: Medium attack, 26 to 50% of the leaf surface, i.e. ½ ;

7: Severe attack, 51 to 75% of the leaf surface, i.e. ¾ :

9: Strong attack, 76 to 100% of the leaf surface, i.e. 4/4.

- <u>Case of BBTD</u>

0 No symptoms ;

1. Streaks on leaves ;
2. Grooves down to the pseudostem ;
3. Discolouration of leaves of normal size ;
4. Reduced size of discoloured leaves ;
5. Bunchy top" appearance

CHAPTER V: SOME SPECIAL FARMING TECHNIQUES

5.0 Introduction

In this chapter we will discuss some techniques for the mass production of plant propagation material to solve the problem of the quantity of material at a given time, and to solve or mitigate the spread of infected material in the context of large-scale farming. Agricultural techniques that can contribute to the vertical green revolution to increase production without necessarily increasing the surface area will also be studied (for example irrigated rice growing and many other methods).

5.2 De BANANIER PIF method.

There are various traditional or improved methods for producing seedlings, but their yield remains modest due to the use of ballooned shoots (MBENGA, 2022). To increase the rate of multiplication we do :

a) Multiplication by decapitation of pseudotrunks.
b) Folding of the pseudo-trunk.
c) Fake decapitation of pseudotrunks.
d) Ridging.

Growers planting small areas of plantain have difficulty obtaining supplies of quality plants (scarcity of plants, high cost, unavailable quantities to buy). To address these concerns, as well as agronomic issues (short fallow periods due to lack of land)

The PIF method (plants from stem fragments) for mass propagation of banana plants was developed by the Centre Africain de Recherche sur le Bananier Plantain (C.A.R.B.A.P.) because of the low production rate of shoots and their poor sanitary quality (presence of nematodes and weevils). The results have been more than satisfactory, since the plant can be grown intensively with high productivity (an average of around fifty plants per bulb) and healthy shoots can be produced in 3 to 4 months, at any time of year (thanks to greenhouse growing, which increases the temperature in winter) (C. MARSAUDON, 2019).

Benefits of the method :

- Reducing the risk of pests spreading (nematodes, weevils),
- Mass production of waste in a healthy environment, at any time of year.
- Homogeneity of plants,
- Innovative device, accessible to all and requiring little technical expertise

Technique: The PIF method involves a number of conventional steps, including

- **Greenhouse construction: this increases the temperature to speed up sprouting. If it is an earthen environment, a tarpaulin should be placed underneath.**

Figure 16: Greenhouse for the PIF method Source (:Mbenga,2022)

- **Substrates to use**: sawdust, rice husks, embers, etc.

Figure 17: Sawdust (one of the substrates for the PIF method)

➤ **Preparation of plant material** :

- Remove the roots and soil around the shoot,
- Wash waste with clean water;
- Leave them to dry for a day on a surface cleaned with bleach.

➤ **Preparation of explants**

- Blank" trimming or dressing the bulbs and washing them in clean water. The advantage of blank trimming is that the plant material is free from nematodes and weevils.

- Trim the sheaths or shell them. This operation produces an "explant",
- Remove the apical bud (make a cross-shaped incision 3 cm deep in the centre of the bulb),
- Treat with insecticide/ematicide/wood ash or an approved product,
- Leave the explants to rest on a clean surface (48h to 72h).

Figure 18: Parging of the source discharge (Mbenga, 2022)

- Fill the germinator to the brim with sawdust or other pre-treated substrate.
- Place the explants in rows until the block is completely full, so that you can see where each one is located,
- Press the explants into the sawdust and cover them with it.
- Close the germinator hermetically then install the shade cloth,
- Water the germinator 2 to 3 times a week, half an hour per session. Watering lowers the temperature in the germinator, moistens the substrate and waters the plants.

➢ **Weaning seedlings and setting up in the nursery**

- 15 days to 3 weeks after planting, check the growth of the plants,
- 40 to 45 days after planting (stage: 2-3 visible leaves), wean the most vigorous plants, using a sharp, disinfected object to separate the young plants from the explant.

Figure 19: Weaning of rejects

- Re-pot these plants in a mixture of normal and enriched potting soil, then water the plants. This action encourages the root system to develop.

  Place the young plants in optimal conditions (away from the sun, watering every day) to limit losses for up to 3 months.

Figure 20: Banana nursery using the PIF method.

5.2 Rapid multiplication of pineapple shoots *(vivo pineapple plant techniques)*

Ananera plantations are usually created from seedlings taken from old plantations.

The supply of pineapple rejects is not always easy for those who want to grow pineapples in large areas. When they are available, they are often of poor quality. Hence the need for growers to produce and package their own if they want to achieve good results in the field. The aim of this technical sheet is to show growers how to use the different methods of propagating pineapple shoots, and how to prepare them properly for planting. (Modeste BIDIMA)

The chronic shortage of shoots in Côte d'Ivoire has created a permanent need for seeds, which limits the extension of plantations and the productivity of the crop. To meet the huge need for planting material, a method of rapid in vivo propagation of shoots from hulled stems has been developed.

5.2.1 Traditional techniques for increasing waste production :

- **The direct stump maintenance method after fruit harvesting**

Generally speaking, there are more than 5 types of discharge:

- **The crown: located** at the top of the fruit in a dormant state which, once planted, resumes its development.
- **The stump cayeu:** which originates on the underground part of the stem or on the collar of the plant. It emits roots that penetrate the soil and generally has longer leaves. This basic cayeu is often called a "shoot" or "sucker".
- **The bulbil:** which starts at the base of the fruit and develops from an axillary bud on the stalk. The bulblet is the best plant material for pineapple propagation because it ensures a very regular cycle. However, bulblets are rare: usually only 1 per plant.
- **The Happe: an** intermediate shoot between the cayeu and the bulbil. It develops from the axillary bud, located at the junction of the stem and the fruit stalk.
- **The stem cayu:** which originates from the axillary bud of the stem. Its terminal bud is similar to the stem. It is this bud that ensures the second harvest on the same plant. Its base has the typical appearance of a duck's beak. The cayeu is most commonly used as plant material in industrial crops.

The stump maintenance method consists of maintaining the stumps in situ, after harvesting the fruit. The plot must be kept clean and aerated to allow the stump to grow. This is how it is done:

- After harvesting, leave the stump to rest for a few days;
- Prune the leaves to reduce the surface area on which nutrients are consumed by the stump, thus speeding up its development;
- Regular weeding of the plot to aerate the stumps and reduce humidity;
- Spray with a mixture of insecticide and fungicide after 3 weeks;
- Add 5g of urea per strain every 3 months.
- Harvest the ripe shoots each month for planting.

This means you can maintain your stumps for 18 months, harvesting 4 to 5 times from each stump, i.e. 4 to 5 shoots per stump.

5.2.2 Method for vivo plants

Multiplication of pineapple shoots on shelled stems is based on old plants or mother plants from a plantation that has already been harvested. The uprooted plants are stripped of their leaves, roots, terminal base and peduncle. The stem obtained is washed in water and cut longitudinally in two.

Preparing growing substrates

Several substrates are used: potting soil, peat, well-decomposed coffee parchment, potting soil + coarse coconut fibre, peat, humus. The different substrate components are mixed in equal portions. Each sectioned part of the stems was lightly pressed into the substrate. The stems are watered

Driving :

- Disinfect the fragments obtained by soaking them completely in a fungicide solution (Ridomil+; 1 sachet in 5 litres of water).
- Make a 15 cm high board with rich black soil, on which you will dig parallel furrows 5 cm apart, inside which the fragments will be planted - Lay the fragments end to end and flat in the furrows, at intervals of 15 cm - Make sure that the cut side is facing the ground, which will help the roots to grow.
- Cover with a thin layer of soil (1 cm) and mulch lightly.
- Water once every 3 days

After three weeks, each fragment will have budded on the upper side and 2 months later, your young seedlings will be 4 to 5 cm high. Set up a nursery consisting of one or more beds 15 to 20 cm high, with shade.

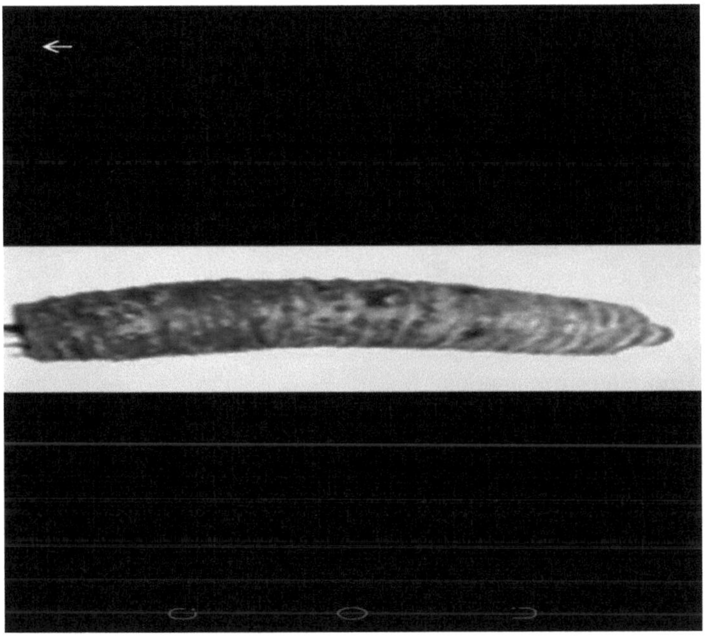

Figure 21: Pineapple stem prepared for the "vivoplant" technique

Conditioning the reject involves draining, grading, disinfecting and trimming.

Rapid yam multiplication

Yam cultivation poses the same problem of propagation material: in traditional cultivation, farmers use the part known as the *"head",* in which case the multiplication coefficient is often slightly higher than 1. To remedy this situation, the technique of rapid multiplication via tubers and multiplication using cuttings from aerial stems *(vine cutting) is* used.

Figure 22: Yam micropropagation.

NB: the technique of rapid multiplication of propagation material is used for all plants with tubers or underground stems (Taro, sweet potato, macabo, etc.).

5.4. Irrigated rice

This is a very important system, as it is of interest to most of the rural population in the Far East (China, South Korea, Japan, Taiwan, North Vietnam, Thailand, India, Sri Lanka).

However, the distribution of this type of farming is limited, particularly where the terrain is flat, there is a lot of water, the soil is very retentive of water or the subsoil is impermeable.

This system delivers high, sustainable yields on the same plot of land. This is due in part to the following factors:

- Keeping the soil under water reduces erosion;
- The nature of the soil limits the leaching of nutrients;
- Raw water and irrigation water bring silt in suspension and other mineral substances in solution;
- The water in rice paddies contains blue-green algae that fix atmospheric nitrogen.

Irrigated rice cultivation has many variations

However, there are a number of points in common:

- The fields are scattered and small in size
- The work is done by members of the family;

- Like shifting cultivation, this is a type of subsistence farming;
- Found in regions with high population densities (Ganges and Tonkin deltas, certain regions of Java: > 1,500/km²; Bangladesh: > 340/km²);
- Rice is the dominant crop;
- Intensive use of land ;
- A lot of work.

The date of arrival and withdrawal, as well as the height of the water wave, are controlled (55% of areas). This is the basis of the green revolution of the 1960s: the simultaneous use of potentially very productive semi-dwarf varieties, mineral fertilisers and pesticides, combined with good weed control through transplanting and manual weeding, enabled average yields of 4 to 5 t/ha to be achieved. Rice is often grown as a single crop. The use of early, non-photosensitive varieties allows up to three cropping cycles per year. With the rise in labour costs, the trend is to abandon transplanting in favour of direct seeding. The major challenge is to improve production levels using techniques that are more respectful of the environment and more economical with water.

➢ **Flooded rice fields**

Here, neither the arrival and withdrawal dates nor the height of the water level are controlled. The most widespread cultivation system is direct seeding. Yields rarely exceed 4 t/ha. The main concern is to keep yields stable at around 3 t/ha. The varieties used must be hardy, and their height and cycle well adapted to the water regime.
A distinction is made between submergence of 0 to 50 cm (23% of areas) and submergence of more than 50 cm, including floating rice (10% of areas).

➢ **Setting up the crop**

There are two main methods: transplanting and direct seeding; the latter can be subdivided according to the condition of the soil at the time of sowing and according to water management after sowing.

- **Soil preparation**

One or two ploughings and several harrows in dry soil or after watering when the crop is planted on mud. Ploughing at the end of the cycle is recommended to bury crop residues and aerate the soil. The seedbed should be prepared or mudded just before sowing or transplanting, to leave the soil free of weeds. These operations can be carried out using mechanised cultivation as well as harnessed or manual cultivation. In rainfed cultivation on slash and burn or plant cover, the soil is not worked.

- **Nursery**

It consists of ensuring the first phase of rice development in a well-controlled environment. The soil is loosened on small planks at , cleared of weeds, smoked (in particular with $P\ O_{25}$), muddied and levelled. Seeds

Pre-sorted and treated with a fungicide-insecticide mixture, pre-sprouted or not, they are sown at a rate of 10 to 20 kg per 100 m2 of nursery. You need 400 m^2 of nursery for one hectare of rice field, a ratio of 1 to 25.

The nursery is submerged after emergence, following the development of the rice, but not exceeding 10 cm. A particularly sophisticated method using plastic film, the *dapog* nursery, has been developed in the Philippines. This reduces the area and quantity of seed used. Its use requires perfectly level rice fields (MONZENGA,2O2O).

Figure 23: Rice nursery.

- **Transplanting**

The most important parameter is the age of the plants at transplanting. The optimum is twenty to thirty days; beyond this there is a negative correlation between the age of the plants and production potential. The plants to be transplanted are uprooted, washed, bundled, dressed and transplanted the same day or within two days at the most. Transplanting should take place at a depth of 2 to 5 cm (any deeper will reduce tillering capacity) and the number of shoots per clump should be between three and ten, depending on the fertility of the soil, which encourages tillering, and the age of the plants, which reduces tillering. Recovery takes place five to fifteen days later, depending on the age of the plants when transplanted.

Figure 24: Replanting rice

- **Direct seeding**

Three different methods are used depending on the water status of the soil at the time of sowing: sowing on mud, sowing in 5-10 cm of water and dry sowing. The first two only apply to irrigated rice growing, while the third is used for intensive irrigated rice growing (Australia, United States, Europe), flooded rice growing (Africa, Asia) and rainfed rice growing. Direct seeding is most often broadcast. In no-till systems, however, it is carried out in patches.

Compared with transplanting, direct seeding has the advantage of being more labour-saving at the start of the cycle, of not subjecting young plants to a physiological shock that lengthens the cycle, and of being more suitable for mechanised cultivation. Its major disadvantages are that it requires more seed and, above all, causes greater weed pressure, requiring more mechanical or chemical weeding.

- **Maintenance**

Irrigation

The water table is a weed control tool, a thermal flywheel, a pH regulator and a regulator of rice growth and development. In general, it is gradually increased as the rice develops, then stabilised at a height of 10 to 25 cm until flowering. As the rice matures, it is gradually dried out, which is important for grain quality. More sophisticated irrigation techniques are available to increase water efficiency. Depending on the type of soil, the length of the rice cycle and the irrigation methods used, water efficiency varies from 0.2 to 1.2 grams of paddy per litre of water consumed.

- **Weed control**

Weeds are often the main limiting factor in rice production. Preventive measures are rarely enough: weed-free seed, cleaning, etc.

Channels and bunds, good soil preparation, judicious use of crop rotations and good water control.

Chemical weed control methods (pre-emergence or post-emergence, with or without a water gap) must be adapted to the way the crop is planted (transplanting or direct seeding), the level of water control and the types of weeds present. The choice of products must also take account of environmental risks, including the risk of weeds developing resistance.

Figure 25: Intensive rice-growing system

5.5 Ratooning

This technique consists of multiplying the regrowth more generally after the first harvest.

Figure 26: Ratoons.

BIBLIOGRAPHY

ANONYMOUS, SD: Agronomists' techniques, farmers' practices.

BIDIMA, M.R., S.D: pineapple cultivation "multiplication and management of sprouts".

CONKLIN, H. C. 1957: Hanunoo agriculture: a report on an integral system of shifting cultivation in the Philippines, No. 12. Rome.

DELESS, T ; ISSALI, E. F.; AUGUSTE, E. ; KOUASSI K.,:2015 : How to mass produce pineapple vivoplants .

DIEDHIOU, P.C.C., 2019 : Etude comparative des rendements et de la rentabilité du Système de Riziculture Intensif (SRI) et du Système Traditionnel dans le département de Ziguinchor

DOUNIAS E., SD: La Diversité Des Agricultures Itinérantes Sur Brûlis (The Diversity of Itinerant Slash-and-burn Farming)

FAO 2015: Healthy soils are the basis of food agriculture.

GREEN, J., 2007: Guide de la formation sur la conduite de la riziculture. Japan International Cooperation Agency.

https://www.province-nord.nc

LITUCHA, J., 2019: General and comparative agriculture. Unpublished courses IFA-YANGAMBI.

MARSAUDON, C., 2019: Planting techniques from stem fragments. Province Nord.

MAZINGA, K, M., 2019: General agriculture course notes. In vitro culture laboratory - FSA-UNILU

MBENGA, Y., 2022: Influence

MONZENGA, J.C., 2017: Notes de cours de phytotechnie générale. Cours inédit IFA- YANGAMBI.

NORMAN, 1975 : tropical cropping systems

MIX
Papier aus verantwortungsvollen Quellen
Paper from responsible sources
FSC® C105338

Printed by Books on Demand GmbH, Norderstedt / Germany